A background to Cornish, an occasional series, number one

PLACE-NAMES IN CORNWALL
HENWYN-TYLLERYOW YN KERNOW

A preliminary list of recommended Cornish language forms of place-names in Cornwall.

Ken George
Pol Hodge
Julyan Holmes
Graham Sandercock

Kesva an Taves Kernewek
The Cornish Language Board

ISBN 0 907064 62 0

HENWYN-TYLLERYOW YN KERNOW

PLACE-NAMES IN CORNWALL

Following research into place names over many years and the resurgence of interest and growth of use of the Cornish language, it is appropriate that a list of Cornish language versions of Cornish place-names should be produced. This is the purpose of this present document which has emerged to a large extent as the result of demand from Cornish speakers and after lengthy discussion.

It should be stressed that the study of place-names is an area of potential confusion for those without a clear focus; this list does *not* seek to dictate what people must use; it does *not* intend to replace familiar forms of Cornish names, but to enhance and supplement them. The following list recommends the most appropriate forms *for those persons writing or speaking the Cornish language*, based firmly and rigorously on research and our present knowledge. It is recognised that some suggestions may prove contentious. However it is only right that Cornish users should have a basic list with which to work as, for example, do users of other Celtic languages whose place-names have also been subject to lesser status in a bilingual situation. The objective of this publication is to remedy the position for users of Cornish.

The basis of the list is the work done by Oliver Padel, formerly of the Institute of Cornish Studies, and by members of the Cornish Language Board into the origins, development and meanings of Cornish place-names in the context of the historical evolution of the language throughout the whole of Cornwall.

As in all bilingual situations, both languages have their own form of the same name, often in an approximate equivalent form. For example English has its own forms, spellings and pronunciation of many place-names from other languages:

English	Cairo	*Arabic*	Al Kahira
English	Moscow	*Russian*	Moskva
English	Lisbon	*Portuguese*	Lisboa
English	Copenhagen	*Danish*	København
English	Cardiff	*Welsh*	Caerdydd
English	Quimper	*Breton*	Kemper
English	Bodmin	*Cornish*	Bosvenegh

For Cornish speakers it is appropriate that they should use, wherever possible, the best Cornish version of Cornish place-names. This is the purpose of this document. Other bodies, such as local authorities, the Post Office and businesses may wish to use or at least be aware of the Cornish versions.

The recommended forms vary in type (see key on page 4):

- some are known Cornish words:
  e.g. **Porthlogh** *(Portloe)*, **porth** (cove) + **logh** (inlet);
  Treger *(Tregeare)*, **tre** (settlement) + **ker** (fort);

- others include personal names, either Cornish or English, which may be no longer in general use and may give rise to uncertainty sometimes:
  e.g. **Sen Ostell** *(St Austell)*; **sen** (saint) + name;
  Eglosveryan *(St Buryan)*; **eglos** (church) + name;

- other names are older versions of existing places:
  Breanek *(St Agnes)*; **bre** (hill) + unknown word;
  Eglosros *(Philleigh)*; **eglos** (church) + **ros** (heath);

- a minority of places have only ever had an English name and the English elements are translated for use by Cornish speakers:
  Aberfal *(Falmouth)*; **aber** (river mouth) + name;
  Tollowarn *(Foxhole)*; **toll** (hole) + **lowarn** (fox).

The initial lack of familiarity with the Cornish spellings should not be a cause for alarm. Look instead at similar situations within Britain in Wales and Scotland, together with our neighbours in Ireland, Brittany, Belgium and other bilingual areas. Let us make the most of our rich linguistic heritage.

The place-names included in the list are those which appear on the Ordnance Survey 1:250 000 map of Cornwall, almost 500 names in all. These have been thoroughly examined by Padel (1988) and subsequently listed by George (1992). Many thousands of other Cornish place-names are of course now well researched and will appear in supplementary lists as time goes on; many others still defy clear interpretation.

The layout of the following pages is as follows:

- column 1: the Ordnance Survey map grid reference;
- column 2: the current Ordnance Survey map spelling;
- column 3: the recommended version and Cornish spelling ;
- column 4: an authentication code (George 1992).

The authentication code is as follows:

code letter	meaning
K	*a Cornish element which is attested, understood and re-spelled as necessary according to the principles of Kernewek Kemmyn*
M	*an obscure element, re-spelled according to examples from middle Cornish and the principles of Kernewek Kemmyn*
O	*an old translation of a non-Cornish element*
N	*a new translation of a non-Cornish element*
H	*a Celtic personal name*
P	*a non-Celtic personal name*
S	*an obscure element rendered in Kernewek Kemmyn according to sound rather than meaning*
A	*an additional element*

Lower case letters are used for formalized equivalents, for phonetic renderings of English personal names and in other cases of doubt.

Ken George, Pol Hodge, Julyan Holmes, Graham Sandercock July 1996

bibliography[1] :

BANNISTER, J. (1871) *Glossary of Cornish Names,* Williams and Norgate, London.

DEXTER, T.F.G. (1926) *Cornish Names*, Longmans Green, London.

PADEL, O.J. (1985) *Cornish Place-name Elements*, Nottingham.

PADEL, O.J. (1988) *A Popular Dictionary of Cornish Place-names,* Alison Hodge, Penzance.

GEORGE, K.J. (1993) *Gerlyver Kernewek Kemmyn,* Kesva an Taves Kernewek, Bosprenn.

GEORGE, K.J. (1991) *Gerlyver Servadow,* Kesva an Taves Kernewek, Bosprenn.

GEORGE, K.J. (1986) *The Pronunciation and Spelling of Revived Cornish,* Kesva an Taves Kernewek, Bosprenn.

GEORGE, K.J. (1992) `Towards a definitive list of settlement-names in Cornwall'. in *Celtic Languages and Celtic Peoples,* ed. C.J. Byrne, M. Harry, P.Ó Siadhail pp. 267-286.

HODGE, P.B. (1994-) *Koloven Henwyn Tylleryow Kernewek,* in *An Gannas,* Liskeard.

HODGE, P.B. and GEORGE K.J. (1992) *Cornish forms of Cornish names* (unpublished list), Grampound Road.

HOLMES, J.G. (1983) *A Thousand Cornish Place-names,* Dyllansow Truran, Redruth.

IVEY, W.F. (1976) *Dictionary of Cornish Dialect,* Helston

NANCE, R.M. (1978) *An English-Cornish and Cornish-English Dictionary,* Kesva an Taves Kernewek, Penzance.

NANCE, R.M. (1955) *A Guide to Cornish Place-names,* Kesva an Taves Kernewek, Penzance.

NANCE, R.M. (1963) *A Glossary of Cornish Sea Words,* Federation of Old Cornwall Socities, Marazion.

POOL, P.A.S. (1985) *The Place-names of West Penwith,* Federation of Old Cornwall Societies, Penzance.

PRYCE, W. (1790) *Archaelogia Cornu-Britannica.*

WHITE, G.P. (1984) *A Handbook of Cornish family Names* (Enlarged Edition), Dyllansow Truran, Redruth.

---

[1] The changing orthography of Cornish over the years together with new sources and research mean that there is some conflicting evidence apparent in the above sources. Those wishing to have a more detailed background to any name should write to the Cornish Language Board's **Place Name Database** at Fentenwynn, Top Hill, Grampound Road, near Truro, Cornwall ☎ 01726 882681.

O.S. G.R.	Name as found on maps today	Recommended Cornish form	(see key)
423703	Albaston	Trevalba	np
223813	Altarnun	Alternonn	KH
399546	Antony	Trevanta	AP
603285	Ashton	Trevonnenn	nN
773433	Baldhu	Baldu	KK
633383	Barripper	Argelteg	NN
285747	Bathpool	Pollbath	NN
966618	Belowda	Boslowsa	KH
297921	Bennacott	Chibynna	nP
659230	Berepper	Argelteg	NN
020600	Bilberry	Krugbylla	NN
777414	Bissoe	Besow	K
735460	Blackwater	Dowrdu	NN
180581	Bocaddon	Boskaswynn	KH
146605	Boconnoc	Boskennek	KH
994734	Bodieve	Bosyuv	KK
073670	Bodmin	Bosvenegh	Kk
861323	Bohortha	Buorthow	K
392347	Bojewyan	Bosuyon	Kh
764533	Bolingey	Melinji	KK
370304	Bosavern	Bosavarn	Kh
098907	Boscastle	Kastell Boterel	NP
035531	Boscoppa	Boskoppa	Kp
441346	Boskednan	Boskennon	Kh
066887	Bossiney	Boskyni	KH
368328	Botallack	Bostalek	KK
404284	Botusfleming	Bosflumyes	KM
320920	Boyton	Trevoya	nP
162621	Braddock	Brodhek	M
403284	Brane	Bosvran	KH
331745	Bray Shop	Shoppa Bre	NS
266911	Brazacott	Chibrosya	nP
617285	Breage	Eglosvreg	KH

904544	Brighton	Trewolow	nn
208065	Bude	Bud	M
783321	Budock Water	Dowr Budhek	NH
015589	Bugle	Karnrosveur	KKK
997701	Burlawn	Boslowen	Kk
722145	Cadgwith	Porthkaswydh	KK
776504	Callestick	Kellestek	M
359697	Callington	Kelliwik	D
435687	Calstock	Kalstok	M
645401	Camborne	Kammbronn	KK
105836	Camelford	Ryskammel	Nk
533352	Canons Town	Tre an Chenon	NnN
223917	Canworthy Water	Boskarn	nK
529388	Carbis Bay	Karrbons	KK
789382	Carclew	Kruglew	KM
123687	Cardinham	Kardhinan	KK
436626	Cargreen	Karrekreun	KK
732413	Carharrack	Karardhek	Kk
790590	Carines	Krow orth Ynys	KKK
849540	Carland Cross	Krows Korlann	NK
678413	Carn Brea	Karnbre	KK
913385	Carne	Karn	K
616374	Carnhell Green	Pras Karnell	nK
799404	Carnon Downs	Goen Garnynn	OK
375533	Carnyorth	Karnyorgh	KK
005557	Carthew	Kardhu	KK
433503	Cawsand	Porthbugh	nN
751444	Chacewater	Dowr an Chas	NNK
997755	Chapel Amble	Amaleglos	KK
038517	Charlestown	Porthmeur	KK
416721	Chilsworthy	Boschyl	np
479310	Chyandour	Chi an Dowr	KKK
180783	Codda	Stummkodda	KM
868613	Colan	Kolan	H
240693	Commonmoor	Goengemmyn	NN
593392	Connor Downs	Goen Gonor	NH
732291	Constantine	Lanngostentin	KH

210116	Coombe	Komm	N
950513	Coombe	Komm	N
209005	Coppathorne	An Spernenn	AN
423685	Cotehele House	Koesheyl	KK
782183	Coverack	Porthkovrek	KM
143967	Crackington Haven	Porthkrag	NK
366543	Crafthole	Toll an Kroft	NAN
791603	Crantock	Lanngorrow	KK
935473	Creed	Krida	H
501367	Cripples Ease	Powes an Mans	NAN
646344	Crowan	Egloskrowenn	KH
516332	Crowlas	Krowrys	KK
902765	Crugmeer	Krugmeur	KK
786578	Cubert	Lannowynn	KM
677213	Cury	Egloskuri	KH
259694	Darite	Darith	M
151873	Davidstow	Lanndhewi	NH
069840	Delabole	Delyowboll	KK
795392	Devoran	Devryon	K
214650	Dobwalls	Fos an Mogh	NAN
199649	Doublebois	Dewgoes	NN
315539	Downderry	Downderri	S
370725	Downgate	Porth an Woen	NAN
439285	Drift	An Drev	KK
234585	Duloe	Dewlogh	KK
256533	East Looe	Logh	K
183635	East Taphouse	Diwotti Est	nn
252159	Eastcott	Chi an Est	nn
001719	Egloshayle	Eglosheyl	KK
272867	Egloskerry	Egloskeri	KH
366270	Escalls	Askall	K
810326	Falmouth	Aberfal	NM
826385	Feock	Lannfiek	KH
809339	Flushing	Nanskersi	Kkk
690386	Four Lanes	Peder Bownder	NN
125517	Fowey	Fowydh	kK
965548	Foxhole	Tollowarn	NN

913583	Fraddan	Frodan	K
398521	Freathy	Frethi	S
703238	Garras	Garros	K
483363	Georgia	Gorga	K
585294	Germoe	Germow	H
873352	Gerrans	Gerens	H
608314	Godolphin Cross	Krows Wodolghan	NM
122548	Golant	Goelnans	KK
544306	Goldsithney	Goelsydhni	KH
730498	Goonbell	Goenbell	KK
788537	Goonhavern	Goenhavar	KK
230164	Gooseham	Pras an Woedh	NAN
013416	Gorran Haven	Porthyust	KH
935485	Grampound	Ponsmeur	OO
916505	Grampound Road	Fordh Ponsmeur	NOO
414336	Great Bosullow	Boswolow Veur	KKK
263070	Grimscott	Chigrymm	nP
485317	Gulval	Lannystli	KM
432717	Gunnislake	Lynngonna	NP
706267	Gweek	Gwig	K
739402	Gwennap	Lannwenep	KH
595374	Gwinear	Sen Gwynnyer	AH
586412	Gwithian	Sen Goedhyan	AH
180878	Hallworthy	Halworgi	KH
507387	Halse Town	Trehals	NP
398697	Harrowbarrow	Kelliloes	NN
349699	Haye	An Hay	Ak
559371	Hayle	Heyl	K
462313	Heamoor	An Hay	KK
758261	Helford	Pennkestell	KK
076710	Helland	Hellann	KK
658275	Helston	Hellys	KK
088813	Helstone	Hellys	KK
267734	Henwood	Koes an Yar	NNN
224604	Herodsfoot	Nanshiryarth	NKK
307574	Hessenford	Rys an Gwraghes	NNN
968532	High Street	Stretughel	NN

768589	Holywell	Porthheylynn	KK
671440	Illogan	Egloshal	KK
919590	Indian Queens	Krows Karworgi	KKH
198958	Jacobstow	Lannjago	NH
808427	Kea	Sen Ke	AH
622410	Kehelland	Kellihellann	KKK
359715	Kellybray	Kellivre	KK
371299	Kelynack	Kelynnek	K
820458	Kenwyn	Keynwynn	KK
443271	Kerris	Kerys	K
851593	Kestle Mill	Melin Gestell	NK
253113	Kilkhampton	Tregylgh	nK
434505	Kingsand	Porthmyghtern	nN
893509	Ladock	Egloslasek	KH
446245	Lamorna	Nansmornow	KM
878417	Lamorran	Lannvorenn	KH
343250	Lands End	Penn an Wlas	OOO
711126	Landewednack	Lanndewynnek	KH
374605	Landrake	Lannergh	K
431615	Landulph	Lanndhelek	KH
228840	Laneast	Lannast	KM
299865	Langore	Nansgover	KK
085636	Lanhydrock	Lannhydrek	KH
039642	Lanivet	Lannives	KK
080590	Lanlivery	Lannlivri	KH
716400	Lanner	Lannergh	K
181569	Lanreath	Lannreydhow	KH
173516	Lansallos	Lannsalwys	KM
147536	Lanteglos Highway	Fordh Nanseglos	NKK
408735	Latchley	Kelligors	NN
244057	Launcells	Lannseles	KM
332846	Launceston	Lannstefan	KH
355824	Lawhitton	Nansgwydhenn	KK
604344	Leedstown	Trelids	NN
548377	Lelant	Lannanta	KH
141571	Lerryn	Leryon	M
131904	Lesnewth	Lysnowydh	NN

276807	Lewannick	Lannwenek	KH
174663	Ley	Kelli	N
338791	Lezant	Lannsans	KK
319736	Linkinhorne	Lanngynhorn	KH
253646	Liskeard	Lyskerrys	KM
918722	Little Petherick	Nansfenten	KK
703125	Lizard	Lysardh	KK
032613	Lockengate	Yet an Toll	NN
745342	Long Downs	Goenyowhir	nN
104597	Lostwithiel	Lostwydhyel	KK
389739	Luckett	Chilova	np
505330	Ludgvan	Lusowan	K
052581	Luxulyan	Logsulyen	KH
764341	Mabe	Lannvab	KK
452319	Madron	Eglosvadern	KH
145706	Maidenwell	Fentenvoren	NN
845427	Malpas	Kammdrog	NN
763250	Manaccan	Managhan	M
517525	Marazion	Marghasyow	KK
224037	Marhamchurch	Eglosvarwenn	NP
370584	Markwell	Fentenvargh	Nh
152919	Marshgate	Porth an Hal	NNN
708250	Mawgan	Sen Mowgan	H
703458	Mawla	Mowla	M
787273	Mawnan	Sen Mownan	M
777287	Mawnan Smith	Mownan an Gov	MNN
100513	Menabilly	Menebelli	KK
288628	Menheniot	Mahunyes	Kh
279661	Merrymeet	Merimet	NN
863448	Merther	Eglosverther	KK
015449	Mevagissey	Lannvorek	KK
081788	Michaelstow	Lannvighal	NH
423527	Millbrook	Govermelin	NN
261712	Minions	Menyon	k
860545	Mitchell	Toll an Voren	NAN
744505	Mithian	Mydhyen	M
402353	Morvah	Morvedh	KK

261569	Morval	Morval	M
206153	Morwenstow	Lannvorwenna	NH
781562	Mount	Menydh	N
148680	Mount	Menydh	N
452527	Mount Edgcumbe	Menydh Pennkomm	NNN
716475	Mount Hawk	Menydh Hok	NN
872603	Mountjoy	Meynji	KK
469263	Mousehole	Porthynys	KK
217565	Muchlarnick	Lannerghmeur	KN
678192	Mullion	Eglosvelyan	KH
804363	Mylor Bridge	Ponsnowydh	KK
638322	Nancegollan	Nansigolenn	KK
960561	Nanpean	Nansbyghan	KK
899522	New Mills	Melinnowydh	NN
425316	Newbridge	Hal an Taken	KKM
464285	Newlyn	Lulynn	KK
828563	Newlyn East	Eglosniwlin	KM
458340	Newmill	Melinnowydh	NN
807620	Newquay	Tewynn Pleustri	KM
741232	Newtown	Trenowydh	NN
272767	North Hill	Bre Gledh	NN
282896	North Petherwin	Paderwynn Gledh	HKN
312972	North Tamerton	Tre war Damer	nAK
844417	Old Kea	Lanndege	KH
163908	Otterham	Prasdowrgi	NN
919754	Padstow	Lannwedhenek	KH
075535	Par	Porth	K
464271	Paul	Pawl	H
203550	Pelynt	Plunennys	KH
999404	Penare	Pennardh	KK
383343	Pendeen	Penndin	KK
024794	Pendoggett	Penndewgoes	KKK
710378	Penhalvean	Pennhal Vyghan	KKK
780308	Penjerrick	Pennansseyrik	KK
080565	Penpillick	Pennpellik	KK
813391	Penpoll	Pennpoll	KK
147544	Penpoll	Pennpoll	KK

876708	Penrose	Pennros	KK
788343	Penryn	Pennrynn	KK
291697	Pensilva	Pennsylva	KS
018473	Pentewan	Bentewynn	KK
793614	Pentire	Penntir Vyghan	KKK
936795	Pentireglaze	Penntirglas	KKK
024563	Penwithick	Penn an Wydhek	KKK
472302	Penzance	Pennsans	KK
776386	Perranarworthal	Pyran ar Woethel	HKK
756543	Perranporth	Porthpyran	KH
537294	Perranuthnoe	Pyranudhno	HM
771521	Perranzabuloe	Pyran yn Treth	KAK
566384	Phillack	Felek	H
871395	Philleigh	Eglosros	KK
366643	Pillaton	Trebeulyow	nN
811414	Playing Place	Plen an Gwari	NNN
333893	Polapit Tamar	Polltarow	NN
348570	Polbathic	Pollbardhek	KM
996504	Polgooth	Pollgoedh	KK
093521	Polkerris	Pollkerys	KK
972456	Polmassick	Ponsmasek	KH
207509	Polperro	Porthpyra	KH
263820	Polruan	Porthruan	KH
263820	Polyphant	Pollefans	KK
938789	Polzeath	Pollsygh	KK
758377	Ponsanooth	Pons an Woedh	KKK
694334	Porkellis	Porthkellys	KK
997807	Port Isaac	Porthusek	KM
003808	Portgaverne	Porthgavran	KM
666178	Porth Mellin	Porthmelin	KK
753277	Porth Navas	Porth an Navas	KKK
796232	Porthallow	Porthalow	KM
383226	Porthcurno	Porthkernow	KK
629258	Porthleven	Porthleven	KK
432371	Porthmeor	Porthmeur	KK
960413	Portholland	Porthalan	KM
806218	Porthoustock	Porthewstek	KM

032506	Porthpean	Porthbyghan	KK
692479	Porthtowan	Porthtewynn	KK
938395	Portloe	Porthlogh	KK
970805	Portquin	Porthgwynn	KK
656453	Portreath	Porthtreth	KK
877353	Portscatho	Porthskathow	KK
355538	Portwrinkle	Porthwikkel	KM
223077	Poughill	Fentenvoekka	NN
202994	Poundstock	Tregorlann	nNN
581282	Prah Sands	Porthgwragh	KK
636357	Praze an Beeble	Pras an Bibell	KKK
669162	Predannack Wollas	Predennek Woeles	KK
899477	Probus	Lannbroboes	KH
313647	Quethiock	Gwydhek	S
726338	Rame	Hordh	N
426493	Rame Head	Penn an Hordh	OOO
079611	Redmoor	Halrudh	NN
699420	Redruth	Rysrudh	KK
798560	Rejerrah	Rysworow	KM
566319	Relubbus	Ryslowbes	KM
924570	Retew	Rysdu	KK
294731	Rilla Mill	Melin Ryslegh	NKK
594273	Rinsey	Rynnji	KK
987602	Roche	An Garrek	OO
933756	Rock	Pennmen	KK
777547	Rose	Ros	K
614397	Roseworthy	Rysworhi	KH
096764	Row	Rew	N
894420	Ruan Lanihorn	Lannrihorn	KH
720152	Ruan Minor	Ruan Vyghan	HK
898703	Rumford	Rysledan	NN
013668	Ruthernbridge	Ponsroethan	NK
924605	Ruthvoes	Rudhfos	KK
433587	Saltash	Essa	
420293	Sancreed	Eglossankres	KH
250567	Sandplace	Tewesva	NN
724442	Scorrier	Skorya	S

304544	Seaton	Seythyn	M
357254	Sennen	Sen Senan	H
350263	Sennen Cove	Porthsenan	NH
707310	Seworgan	Ryswoedhgen	KH
370551	Sheviock	Seviek	K
877735	Shop	Shoppa Parkyn	NH
228148	Shop	Shoppa	N
808477	Shortlanesend	Penn an Vownder	KKK
636289	Sithney	Sen Sydhni	H
330726	South Hill	Bre Dheghow	NN
310818	South Petherwin	Paderwynn Dheghow	HKN
244929	South Wheatley	Kelliwynn Dheghow	nNn
720507	St Agnes	Breanek	KM
823506	St Allen	Eglosalan	KH
416709	St Ann's Chapel	Chapel Anna	NNN
782256	St Anthony	Lannentenin	KH
014524	St Austell	Sen Ostell	NH
069548	St Blazey	Lanndreth	KK
978717	St Breock	Nanssans	KK
097773	St Breward	Sen Branwalader	NH
409257	St Buryan	Eglosveryan	KH
247682	St Cleer	Sen Kler	NP
851438	St Clement	Sen Klemens	NH
205843	St Clether	Sen Kleder	NH
913636	St Columb Major	Sen Kolomm Veur	NPN
840623	St Columb Minor	Sen Kolomm Vyghan	NPN
912636	St Columb Road	Fordh Sen Kolomm	NNP
730426	St Day	Sen Day	NH
950577	St Dennis	Tredhinas	AK
401678	St Dominick	Sen Domynek	NH
997786	St Endellion	Sen Endelyn	NH
893569	St Enoder	Eglosenoder	KH
846499	St Erme	Egloserm	KH
551351	St Erth	Lannudhno	KM
576352	St Erth Praze	Pras	K
892703	St Ervan	Sen Erven	NH
978462	St Ewe	Lannewa	KH

159973	St Gennys	Sen Gwynnys	NH
359577	St Germans	Lannales	KM
551313	St Hilary	Bronnlowena	KK
928718	St Issey	Egloskrug	KK
309672	St Ive	Sen Iv	NP
518405	St Ives	Porthia	KH
408537	St John	Sen Yowann	NP
859357	St Just in Roseland	Lannsiek	KH
370315	St Just in Penwith	Lannyust	KH
790213	St Keverne	Lannaghevran	KH
022768	St Kew	Lanndogho	KH
033755	St Kew Highway	Fordh Lanndogho	NKH
242609	St Keyne	Sen Keyn	NH
037663	St Lawrence	Sen Lorens	NH
380223	St Levan	Sen Selevan	NH
042732	St Mabyn	Sen Mabon	NH
260550	St Martin	Sen Martin	NH
847331	St Mawes	Lannvowsedh	KH
873660	St Mawgan	Sen Mowgan	NH
388655	St Mellion	Sen Melyan	NH
963420	St Michael Caerhayes	Lannvighal	KH
858422	St Michael Penkevil	Pennkevyll	KK
966771	St Minver	Sen Menvra	NH
200632	St Pinnock	Sen Pynnek	NH
951504	St Stephen Coombe	Komm Sen Stefan	NNP
945523	St Stephen in Brannel	Eglosstefan	KH
325856	St Stephens	Lannstefan Wartha	KPN
417583	St Stephens	Sen Stefan	NP
064806	St Teath	Eglostedha	KH
066763	St Tudy	Eglostudi	KH
968648	St Wenn	Sen Gwenna	NH
115570	St Winnow	Sen Gwynnow	NH
014571	Stenalees	Steneklas	KK
226107	Stibb	Stokk	N
980502	Sticker	Stekkyer	K
735367	St. Stithians	Sen Stedhyans	M
360744	Stoke Climsland	Eglosklym	Np

231064	Stratton	Strasnedh	KK
887562	Summercourt	Marghashir	NN
912653	Talskiddy	Talskisi	N
146734	Temple	Tempel	N
347597	Tideford	Rysteusi	NM
057884	Tintagel	Tre war Venydh	KKK
050890	Tintagel Castle	Dintagell	KK
440550	Torpoint	Penntorr	NK
487381	Towednack	Tewynnek	H
592328	Townshend	Penn an Dre	NNN
743213	Traboe	Treworabo	KM
264773	Trebartha	Trebartha	KM
056860	Trebarwith	Trebervedh	KK
934781	Trebetherick	Trebedrek	KH
895616	Trebudannon	Trebydannan	KM
349777	Treburley	Treborley	KP
454286	Tredavoe	Trewordhavo	KH
927703	Tredinnick	Treredenek	KK
394230	Treen	Tredhin	KK
298838	Tregadillet	Tregadyles	KM
243867	Tregeare	Treger	KK
754230	Tregidden	Tregudynn	KK
193981	Tregole	Tregowl	KM
956639	Tregonetha	Tregenhetho	KH
925448	Tregony	Trerigni	KH
851653	Tregurrian	Treguryan	KS
990792	Trelights	Treleghrys	KKK
044780	Trelill	Trelulla	KP
837396	Trelissick	Trelesik	KH
720239	Trelowarren	Trelowaren	KS
163865	Tremail	Trevel	KH
236890	Tremayne	Treven	KK
257682	Tremar	Trevargh	KH
394597	Trematon	Treveu	KM
853681	Trenance	Trenans	KK
033487	Trenarren	Dingaran	KK
208881	Treneglos	Treneglos	KK

173533	Trenewan	Trenowyen	KM
578309	Trescowe	Treskaw	KK
868464	Tresillian	Tresulyen	KH
233874	Tresmeer	Trewasmeur	KH
038556	Trethurgy	Tredhowrgi	KH
098907	Trevalga	Trevelgi	KH
978728	Trevanson	Trevanson	Km
912603	Trevarren	Treveren	KM
983432	Trevarrick	Trevarek	KM
742521	Trevellas	Trevelys	KM
756317	Treverva	Trevurvo	KM
336697	Trevigro	Trevigra	KM
943568	Treviscoe	Trevosker	KH
893754	Trevone	Treavon	KM
066866	Trewarmett	Treworman	KM
903443	Trewarthenick	Trewedhenek	KH
147868	Trewassa	Trewasso	KH
377338	Trewellard	Trewylardh	KM
253835	Trewen	Trewynn	KK
256598	Trewidland	Trewydhlann	KK
186975	Trewint	Trewyns	KK
879373	Trewithian	Trewydhyan	KH
996527	Trewoon	Trewoen	KK
863737	Treyarnon	Treyarnenn	KH
842503	Trispen	Tredhespan	KM
662382	Troon	Trewoen	KK
827449	Truro	Truru	M
760423	Twelve Heads	Dewdhek Stamp	NN
085543	Tywardreath	Chi war Dreth	KKK
281721	Upton Cross	Krows Trewartha	NnN
916395	Veryan	Elerghi	K
990617	Victoria	Trevudhik	An
990723	Wadebridge	Ponsrys	NN
181955	Wainhouse Corner	Stumm an Gwinji	NNN
207907	Warbstow	Lannwarburgh	NP
156690	Warleggan	Gorlegan	MM
237976	Week St Mary	Gwigvaria	SNN

678310	Wendron	Egloswendron	KH
328877	Werrington	Trewolvrin	np
282937	West Curry	Kori Veur	NN
253533	West Looe	Porthbyghan	KK
776607	West Pentire	Penntir	KK
970573	Whitemoor	Halwynn	NN
265985	Whitstone	Mengwynn	NN
203012	Widemouth	Aswaledan	NN
995654	Withiel	Gwydhyel	K
220135	Woodford	Ryskoes	NN
318874	Yeolmbridge	Ponsyam	NM
827449	Zelah	An Hel	AN
455385	Zennor	Sen Senar	AH

Additional information or background details on the names included here can be obtained through the Cornish Language Board's **Place-name Database**, Top Hill, Fentenwynn, Grampound Road, near Truro, Cornwall ☎.01726 882681.

Also available in this series:

A background to Cornish, number 2,
The Formation of Cornish Place-names
by Graham Sandercock and Wella Brown;

A background to Cornish, number 3,
A Very Brief History of the Cornish Language
by Graham Sandercock.

Those wishing to support the work of the Cornish language can join *Kowethas an Yeth Kernewek*, the Cornish Language Fellowship, which works closely with the Cornish Language Board. The Fellowship produces a monthly Cornish language magazine, *An Gannas*. Details from Trewynn, Lodge Hill, Liskeard, Cornwall.

Other publications from *Kesva an Taves Kernewek* (The Cornish Language Board) include:

Cornish This Way ~ *Holyewgh an Lergh*
by **Graham Sandercock** ISBN 0 907064 12 4, 64 A4 pages, softback,
a course for beginners in Cornish
plus accompanying double cassette
ISBN 0 907064 17 5, three hours running time

Grammar for the First Grade ISBN 0 907064 13 2 and
Grammar Beyond the First Grade ISBN 0 907064 14 0
by **John Page,**
48 A5 pages each

Gerlyver Kernewek Kemmyn by **Dr Ken George**
ISBN 0 907064 11 6, 338 pages, hardback
the major modern scholarly dictionary
Cornish-English with full derivations, notes and
Breton and Welsh cognates

A Grammar of Modern Cornish by **Wella Brown**
ISBN 0 907064 09 4, 253 pages, hardback
a comprehensive and detailed description
of Cornwall's own language - for the advanced student

The First Thousand Words in Cornish ISBN 0 907064 16 7
the Cornish version of this popular and
colourful series for children and adults alike

THESE TITLES, AND MORE, AND DETAILS OF CORNISH COURSES ARE AVAILABLE FROM GEORGE ANSELL, CORNISH LANGUAGE BOARD PUBLICATIONS

65 Churchtown, Gwinear, Hayle, Cornwall
☎ 01736 850878; fax 01736 850878